Education
144

# 是守护人还是小偷

## Watchdog or Thief

### Gunter Pauli

[比]冈特·鲍利 著

[哥伦]凯瑟琳娜·巴赫 绘

何家振 译

上海远东出版社

# 丛书编委会

主　任：田成川

副主任：闫世东　林　玉

委　员：李原原　祝真旭　曾红鹰　靳增江　史国鹏

　　　　梁雅丽　孟小红　郑循如　陈　卫　任泽林

　　　　薛　梅　朱智翔　柳志清　冯　缨　齐晓江

　　　　朱习文　毕春萍　彭　勇

特别感谢以下热心人士对童书工作的支持：

匡志强　宋小华　解　东　厉　云　李　婧　庞英元

李　阳　梁婧婧　刘　丹　冯家宝　熊彩虹　罗淑怡

旷　婉　王靖雯　廖清州　王怡然　王　征　邵　杰

陈强林　陈　果　罗　佳　闫　艳　谢　露　张修博

陈梦竹　刘　灿　李　丹　郭　雯　戴　虹

# 目录

是守护人还是小偷　　4

你知道吗?　　22

想一想　　26

自己动手!　　27

学科知识　　28

情感智慧　　29

艺术　　29

思维拓展　　30

动手能力　　30

故事灵感来自　　31

# Contents

Watchdog or Thief　　4

Did You Know?　　22

Think About It　　26

Do It Yourself!　　27

Academic Knowledge　　28

Emotional Intelligence　　29

The Arts　　29

Systems:
Making the Connections　　30

Capacity to Implement　　30

This Fable Is Inspired by　　31

一只来自卡拉哈里沙漠的狐獴对着一只织巢鸟生气，指责他给了错误的警报信号。

"我刚刚丢了我的食物！因为你的警报，我赶紧去躲避，结果扔掉了刚抓到的那只鲜美多汁的蝎子，没想到附近根本就没有捕食者。"

A meerkat from the Kalahari Desert is angry at a weaver bird, accusing him of giving a false alarm.
"I just lost my meal! I dropped the juicy scorpion I had just caught and ran for cover because of your alarm call, only to realise that there are no predators around at all."

狐獴对着一只织巢鸟生气……

A meerkat is angry at a weaver bird ...

是卷尾……

It was the drongo ...

"不是我。"织巢鸟辩解道，"是卷尾，他偷了你的蝎子。"

"卷尾？难以相信！他是我们的守护人，一直密切留意，有危险的时候会向我们发出警告。"

"哦，那是他想让你信任他。卷尾就是这样一个骗子。"

"That wasn't me," Weaver defends himself. "It was the drongo, he stole your scorpion."
"The drongo? That is hard to believe! He is our watchdog, always on the lookout, warning us of impending danger."
"Oh, that is what he wants you to believe. That drongo is such a liar."

"但是那声音听起来就像织巢鸟的警报信号——响亮而清晰。你才是那个拿走我辛苦弄来的食物的人。"

"多么悲哀呀，明明是那只狡猾的卷尾模仿我的警报信号，你却觉得我是贼。"

"But it sounded just like a weaver alarm call – I heard it loud and clear. You are the one who got my hard-earned meal for free!"

"How sad that you believe I am a thief, when it was that sly drongo – imitating my alarm call."

......狡猾的卷尾模仿我的警报信号。

... sly drongo - imitating my alarm call.

......在空中捕获昆虫的惊人本领。

... captures insects mid-air with incredible skill.

"噢，得了吧，"狐獴回应道，"卷尾有在空中捕获昆虫的惊人本领。他肯定不需要靠欺骗我们来抢夺食物。而你总是忙于织你的巢，所以才没有时间寻找食物。"

　　"你没注意到在寒冷的早晨这附近根本没有昆虫飞过吗？你以为卷尾就只是忍着饥饿吗？"

"Oh come on," Meerkat responds, "the drongo captures insects mid-air with incredible skill. He definitely doesn't need to fool any of us out of our food. You are so busy weaving your nest that you have no time for finding food."

"Have you noticed there are no insects flying around here on cold mornings? Do you think that the drongo will simply go hungry?"

"但很确定他只说卷尾语！你是想说卷尾也会说织巢鸟语吗？"

"卷尾比我们任何人会说的语言都多。他甚至还会说狐獴语。"

"真的吗？我看见他毫不畏惧地赶走了鹰。他肯定不需要靠欺骗来得到食物。"

"But surely he will speak drongo-ese! Are you telling me that he can speak also weaverese?"

"Drongo speaks more languages than any of us. He can even speak meerkatese."

"Really? I have seen how he chases off eagles and hawks, without any fear. Surely he does not have to resort to cheating to get food."

他甚至还会说狐獴语。

He can even speak meerkatese.

指责我这个织巢大师公平吗……

... is it fair to accuse me, the master weaver ...

"听着，我知道你对卷尾评价很高，但是指责我这个织巢大师和鸟巢建设者是骗子就公平吗？把你的指责送给应该被指责的人吧：那只长着红色眼睛、钩状的嘴和叉尾的黑色鸟儿。"

"得了吧，到目前为止他给我的都是可靠的信息。"狐獴说。

"Look, I know you think highly of drongos, but is it fair to accuse me, the master weaver and nest builder, of cheating? Place the blame where it belongs: with that black bird with his red eyes, hooked beak and forked tail."

"Well, so far he has always only given me reliable information," Meerkat says.

"卷尾很了解我们，他知道我们会如何反应。他会说我们织巢鸟的语言，能模仿苍鹰的叫声。相信我，我曾经听过他模仿你们狐獴说话！"

"这周围也许是有很多欺骗行为，我们可能会上当，但是被我们最亲近的看护人卷尾欺骗了？我还是觉得很难相信！"

"That drongo knows us well. He knows how we will react. He can speak my weaver language, and copy the goshawk's call. And believe me, I have heard him imitate your meerkatese!"

"There may be a lot of deception going on around here, and we are getting fooled, but by our very own watchdog, the drongo? I still find that hard to believe!"

我还是觉得很难相信！

I still find that hard to believe!

......欺骗另一种鸟或动物，让它们丢掉食物！

... cheat another bird or animal out of its food!

"卷尾非常聪明，他们能学会大约五十种不同的动物和鸟类的警报声音。"

"哇，你是说他能说五十种语言？"

"是的，他可以用五十种语言欺骗。如果他用一种语言欺骗失败了，他能很容易地转换到另一种语言，欺骗另一种鸟或动物，让它们丢掉食物！"

"Drongos are so clever, they can learn the alarm calls of about fifty different animals and birds."

"Wow, you mean he can speak fifty languages?"

"Yes, he speaks, and cheats, in fifty different languages. So if his deception in one language fails, he can easily switch over to another. And cheat another bird or animal out of its food!"

"既然我知道了这些，我将会密切关注这位我们最钟爱的'守护人'。"

"我希望你不要再指责我这个无辜者了。毕竟我只是一只食籽鸟，我甚至不吃昆虫，当然也不会吃蝎子！"

……这仅仅是开始！……

"Now that I know this, I will be keeping a close eye on our favourite 'watchdog'. "

"And I hope you will no longer accuse me, the innocent one in all of this. I'm a seedeater after all and don't even eat insects and certainly not a scorpion! "

... AND IT HAS ONLY JUST BEGUN! ...

······这仅仅是开始！······

... AND IT HAS ONLY JUST BEGUN! ...

# Did You Know?

## 你知道吗?

In an Australian dialect the word drongo means "idiot", but a bird that can imitate the sounds of so many other birds and animals, should be called anything but an idiot.

在一种澳大利亚方言里，单词"卷尾"的意思是"白痴"，但是一种能够模仿很多鸟类和动物声音的鸟，根本不应该被称为白痴。

Drongos are excellent hunters, catching insects mid-air. They are fearless and will attack other birds that are much larger than them.

卷尾是出色的猎人，在空中抓昆虫。它们无所畏惧，会攻击比它们大很多的鸟类。

卷尾在很多栖息地开创了自己独特的空间，并进化出很多不同的种类。安达曼卷尾、科摩罗卷尾、马约特卷尾和阿岛卷尾，每一个品种都是当地特有的，因而以其所在的岛屿命名。

Drongos have carved out a unique niche for themselves in many habitats and have evolved into many different species. The Andaman Drongo, the Comoro Drongo, the Mayotte Drongo and the Aldabra Drongo, each endemic to the islands with the same name.

织巢鸟有100多个品种。雄性筑巢意在吸引雌性，鸟巢开始是单股，然后打结成簇，最终成为一个安全的窝。

There are over one hundred weaver species. The male builds the nest in the hope of attracting a female. The nest starts with a single strand, knotted to a branch and grows into a safe home.

By taking prey from other animals, drongos are able to obtain food items that are too big for them to catch themselves, such as geckos, and also scorpions – after a meerkat has removed its venomous sting.

卷尾能够从其他动物那里获取那些它们无法自己捕猎到的大猎物，比如壁虎，还有蝎子——在狐獴将蝎子有毒的尾刺剥掉之后。

When the weaver's nest is ready, the male will invite females for an "open house viewing". If a female approves of it, it will mate with the male and lay eggs. If no female likes it, the nest will be abandoned. A male weaver may have to build several nests during the mating season.

如果织巢鸟的巢织好了，雄性会邀请雌性进行"开放式参观"。如果雌性喜欢这个巢，它就会与这只雄性交配并产卵。如果没有雌性喜欢，该巢就被废弃了。在交配季，一只雄性织巢鸟可能不得不建数个巢。

The sparrow weavers in Africa will aggregate nests into a large community of hundreds of pairs of weavers. This communal nesting is similar to people living in an apartment building.

非洲的一种麻雀织巢鸟会把巢穴聚集到一起，构成一个由数百对织巢鸟组成的群落。这种群聚巢穴类似于人类的公寓大楼。

Drongos obtain nearly a quarter of their daily food intake by making false alarms calls, mimicking the sounds made by other birds, species, including babblers, glossy starlings, weavers and goshawk, and even that of the meerkat.

卷尾的日常食物中约有四分之一是通过模仿其他鸟类和动物的声音（包括画眉、辉椋鸟、织巢鸟和苍鹰甚至狐獴的声音），发出假警报信号获得的。

Would you like to learn another language so that people believe you are someone else?

你愿意学习另一种语言，让别人以为你是其他人吗？

Is it acceptable to trick others into believing that there is danger so that you can steal their food?

欺骗别人，让对方相信有危险，以便偷走他们的食物，这是可以接受的吗？

How many times do you think one has to hear a false fire alarm go off, before one no longer pays any attention to it?

你认为一个人听到多少次假的火警警报以后，就不再把警报当回事儿了呢？

Stealing food is wrong, but what if you are very hungry and it is your only option?

偷取食物是错的，但是如果你很饿，而且这是你的唯一选择呢？

# Do It Yourself!

# 自己动手!

Let's look at alarms. How many do you know of? How many of those do you pay attention to? Find out what alarms are required for your safety at school and where you live. How often do you have safety drills? Do you all participate? If you live in an area with earthquake risk, it is important to learn how to protect yourself and your family in case of an earthquake. If you have already taken a course in first aid, so you know what to do in a medical emergency, do so now. Teach others what to do in case of an emergency as well.

我们看看警报吧。你知道警报有多少种吗?有多少警报你注意过?找出为了学校和居住地安全所需的警报。你多久进行一次安全演习?你每次都参加吗?如果你住在一个有地震风险的地区,学会在地震发生时保护自己和家人的安全是很重要的。如果你学习过急救课程,知道在医疗急救时该如何做,现在就去做吧。也教教其他人在紧急情况下应该怎么做。

## 学科知识
### Academic Knowledge

| | |
|---|---|
| 生物学 | 世界各地25种卷尾中，只有非洲卷尾有非常复杂的发音系统；织巢鸟是雀形目鸟，与燕雀是亲戚；20%会唱歌的鸟会模仿其他鸟的声音；所有蝎子中只有少数有能杀死人的毒液。 |
| 化 学 | 从蝎子毒液中提取的毒素被用于生产杀虫剂、疫苗和蛋白质工程支架。 |
| 物 理 | 空气就像水一样是一种流体，虽然它不像水一样处于液态；流体的名称源自使其形变所需的力依赖于形变速度而非形变大小。 |
| 工程学 | 鸟类和飞行昆虫的重量、翅膀的表面积大小，决定了某一种类的飞行能力；升力是流体流经一个物体表面时产生的垂直于流体流向向上的分力。 |
| 经济学 | 盗窃的成本，包括盗窃以及企业（商店和工厂）采取的安全措施，是计算并被加进顾客支付的价格里的，这意味着偷窃的损失由所有顾客买单。 |
| 伦理学 | 生存权高于财产权，根据此原则，划分出当饥饿时为生存而偷窃和与其他偷窃之间的界限；2016年意大利法院做出裁决，偷窃少量食物以避免饥饿不是犯罪；对有些文化来说，拥有超过所需即为偷窃；少偷仍被视为偷窃，而少排放污染却是被允许的，甚至会被赞美，虽然所有人都认为两种行为都有社会危害性。 |
| 历 史 | 已知最古老的蝎子存在于四亿三千万年前；1641年笛卡儿在他的《第一哲学沉思集》中引入了自我欺骗的概念；尼科洛·马基雅维利在他的《君主论》一书中提到用谎言和欺骗保住权力。 |
| 地 理 | 卷尾存在于非洲、亚洲和澳大利亚，在开放的森林、矮树丛和干旱区；蝎子存在于除南极洲以外的世界各地。 |
| 数 学 | 计算机安全领域的欺骗。 |
| 生活方式 | 谎言就是明知那是假的，却歪曲事实；角色扮演可以有效理解那些在精神病院和受刑事监禁的人的生活；匹诺曹的故事教育儿童不要撒谎。 |
| 社会学 | 在占星术时代，巴比伦天文学家确定了天蝎座是黄道十二宫中的一个；现代法律以民事公平而不是刑事惩罚的观点看待盗窃。 |
| 心理学 | 盗窃癖：一种心理状态，在这里盗窃所得的物品不是用于满足生理需要，而是一种心理愿望；欺骗可以表现为疏忽、谎言和伪装；一旦你认定一个人是有罪的，你还会相信对方吗？自欺欺人；马基雅维利说，想要骗人的人会找到愿意被欺骗的人。 |
| 系统论 | 社会需要诚信的氛围，诈骗和盗窃造成的压力可以导致社会崩溃。 |

## 情感智慧
### Emotional Intelligence

**狐獴**

狐獴被欺骗了，抱怨自己丢掉了食物。她指责织巢鸟欺骗了她，并详细解释了发生了什么。织巢鸟说不是他干的，狐獴表示不相信，不接受织巢鸟关于卷尾应该对盗窃负责的解释，仍然认为是织巢鸟为了得到免费食物而欺骗她。狐獴一开始拒绝相信织巢鸟的解释，并指出卷尾不需要偷的事实。但当她得知卷尾能够模仿五十多种鸟和动物的声音后，开始怀疑自己的看法。她现在准备好保持警觉了。

**织巢鸟**

当受到错误的指责时，织巢鸟迅速为自己辩护，并指明了真正的罪魁祸首——卷尾。当狐獴不相信他时，织巢鸟措辞强烈，指责卷尾是说谎的人，然后解释了卷尾怎样伪装自己的声音，模仿其他鸟类和动物的警报。织巢鸟表示被错误地指责为贼是悲哀的。他诉诸逻辑推理：在寒冷的早晨空中没有昆虫可抓，卷尾一定很饿。织巢鸟抛出的这个问题促使狐獴思考。织巢鸟认为，在卷尾偷窃了食物的情况下，指责他这个编织大师和筑巢专家是不公平的。当织巢鸟指出卷尾会模仿至少五十种鸟类和动物的声音后，狐獴相信了他，并表示她将来会仔细观察卷尾的行为。织巢鸟将不再被视为罪犯，他如释重负。

## 艺术
### The Arts

学习织巢鸟筑巢的方法。一切起始于一个简单系在树枝上的植物条带，像一条绳索。其中最关键的是打一个结。我们可以通过观察织巢鸟的做法学会筑巢基本技能。织巢鸟只用它的喙和爪子就能把巢筑好，而对你来说，用双手和十指做同样的事情应该更容易。把你打的结挨着摆在一张纸上，看看有多么漂亮。你已经把技术转化成了艺术。

## 思维拓展
### Systems: Making the Connections

在我们的世界里，欺骗是一种现实。纵观历史，各种欺骗一直被记述。佛罗伦萨政治家、哲学家和外交家马基雅维利在他的《君主论》一书中，详细描述了君主应该如何使用谎言来欺骗。尽管如此，他仍然被认为是政治学的创始人。马基雅维利主张为达目的可以不择手段。欺骗行为无助于我们建立诚信社会。一旦我们形成一种观念，就很难改变，就如同卷尾能够让狐獴相信织巢鸟是那个小偷一样。在我们社会中有很多使用欺骗手段的人，比如设立公司为非法途径获得的资金转化为正常的银行账户资金提供合法庇护。自然界虽然遵循进化的路径，但也存在少量欺骗行为，可是如果人类希望朝着满足所有人基本需要的方向发展，那么我们必须确保欺骗不能作为可接受的生活方式。这往往是困难的，因为一些人不仅欺骗其他人，甚至还会自我欺骗。接着就会产生伦理道德问题：那些因为贫穷需要生存的人，可能会诉诸欺骗以谋生。在此背景下，我们必须意识到，一些人被无端地指责，仅仅是基于信念和感觉，还有一些人的欺骗行为或许的确需要被宽恕。最终生活是一个动态系统，必须不断完善，不要只是指责错误的东西，取而代之的是，我们可能希望鼓励美好，比如，我们可以赞美卷尾会说多种语言的能力，因为它可以利用这种独特的技能为社会做贡献。

## 动手能力
### Capacity to Implement

在你的朋友或家人中，你怀疑谁没有完全讲真话，或者提供假事实，甚至可能自欺欺人？如果发现有人撒谎或欺骗，你会如何处理？你会指责那个人吗？或者，你会试着找出他们这样做的原因，发现他们有哪些优点吗？与那些也关心这个人的人一起关注他或她的长处，想出能促使他或她保持诚实的方法。通过这种方式帮助别人，我们可以共同创造一个更加美好的未来。

## 故事灵感来自

## This Fable Is Inspired by

# 阿曼达·雷德利

# Amanda Ridley

阿曼达·雷德利在林肯大学（新西兰）学习动物生态学，并获得理学学士学位。后来，她在剑桥大学获得行为生态学博士学位。在南非开普敦大学工作几年之后，雷德利博士转入西澳大利亚大学进化生物学中心。她现在是进化生物学家，专注长期种群动态、巢内寄生体在合作系统内的相互作用。她还进行行为生态学和保护生态学的交叉研究，利用行为学信息为保护性决策的制定提供更多帮助。她也在南非卡拉哈里沙漠地区工作过，在那里她对卷尾的行为进行了第一手观察。

图书在版编目（CIP）数据

冈特生态童书.第四辑：修订版：全36册：汉英对照 /
（比）冈特·鲍利著；（哥伦）凯瑟琳娜·巴赫绘；
何家振等译.—上海：上海远东出版社，2023
书名原文：Gunter's Fables
ISBN 978-7-5476-1931-5

Ⅰ.①冈… Ⅱ.①冈… ②凯… ③何… Ⅲ.①生态环
境–环境保护–儿童读物—汉、英 Ⅳ.①X171.1-49

中国国家版本馆CIP数据核字（2023）第120983号
著作权合同登记号图字09-2023-0612号


**策　　划** 张　蓉
**责任编辑** 曹　茜
**封面设计** 魏　来 李　廉




冈特生态童书

**是守护人还是小偷**

［比］冈特·鲍利　著
［哥伦］凯瑟琳娜·巴赫　绘

何家振　译

记得要和身边的小朋友分享环保知识哦！
八喜冰淇淋祝你成为环保小使者！